NURSING AS A PROFESSION

AND

PATIENT LEADING, GUIDANCE, & SUPPORT

Nursing as a Profession and Patient Leading, Guidance, & Support

Shreen Gaber
RN, MScN, BSc
Nursing Administration Department
Faculty of Nursing
Cairo University

Yale | UNIVERSITY PRESS

2016

First Printing: 2016

ISBN: 978-1-365-64312-5

Yale University Press

Yale University Press
 47 Bedford Square
London
p. 020-7079-4900
Yalebooks.co.uk

Yale University Press London distributes to most countries outside of North and South America. Special discounts are available on quantity purchases by corporations, associations, educators, and others. For details, contact the publisher at the above listed address.

U.S. trade bookstores and wholesalers: Please contact Shreen Gaber Tel: (+20)100-8144971; Fax: (+20) 23657190 or email: Sameh17@cu.edu.eg

Shreen Gaber

To my lovely huspend and children

Thank you. Without your support and persistence, I would have

never accomplished this work.

Content

Acknowledgements

I would like to express my great thanks and appreciation to my family, parents, colleagues, teachers and my students who are always willing to provide their support and guidance. I also appreciate the efforts of (Yale University Press) acquisitions editors and there guidance.

Shreen Gaber

Preface

This book draws an attention to the approaches of patient leading, guidance, & support as one of professional nursing roles and describe each approach's benefits and consequences and how to achieve. In addition, to provide the suggested guidelines, strategies that help the patients to be more cooperative. The researcher strives to write the book in comprehensive, concise and feasible sequential steps to be easily understood aspiring at help the professional nurses to apply the mentioned approaches and improve the quality of care. Whether you are a beginner or an experienced health care provider, I hope that you find this book enjoyable and clinically.

Nursing as a Profession and Patient Leading, Guidance, & Support

Introduction:

Nurses work to promote health, prevent disease and help patients cope with illness. They are advocates and health educators for patients, families and communities. When providing direct patient care, they observe, assess and record patient symptoms, reactions and progress. Nurses collaborate with physicians in the performance of treatments and examinations, the administration of medications and the provision of direct patient care in convalescence and rehabilitation. This is why nursing considered as a profession.

Patient guidance has been an essential component of Registered Nurses' professional role. In today's healthcare delivery system, patient guidance and leading is becoming increasingly important. The public demands more information and control over their health care. Moreover, illness and limited knowledge about medicine, nursing and the healthcare system mean that patients are often vulnerable and powerless. Furthermore, the development of medicine and advanced technology has resulted in aggressive and swift use of technologies, which is becoming the norm in the contemporary hospital environment.

Nursing as a Profession Career

Nurses work in an environment that is constantly changing to provide the best possible care for patients. They are continuously learning about the latest technology and medication as well as considering the evidence that their nursing practice is based upon. Because they will actually spend more face-to-face time with a patient than doctors, nurses must be particularly skilled at interacting with patients, putting them at ease, and assisting them in their recovery. It is often said that physicians cure, and nurses care.

Nurses aiming at promote health, prevent disease and help patients cope with illness. They are advocates and health educators for patients, families and communities. When providing direct patient care, they observe, assess and record patient symptoms, reactions and progress. Nurses collaborate with physicians in the performance of treatments and examinations, the administration of medications and the provision of direct patient care in convalescence and rehabilitation.

Recently there are many articles on the web where the nurse author's stated intent was to "enlighten" future and prospective nurses to the "harsh realities" of the profession. The piece listed things like the physicality of the job, the necessity of doing shift work, and a proclamation that nursing is not a *profession* but just a *job*. Holy encephalopathy, Batman! Are we still having this conversation?

I'm not going to bore you with definitions from Webster's dictionary. Nor am I going to quote the many research papers on this subject from various scientific disciplines. I'm neither a nurse researcher nor a social scientist. I am, however a nurse who in her 35 years in the *profession* has a pretty good idea of what nursing is and what it isn't. When you come right down to it, the following is all the evidence I or anyone else needs to put the issue to rest. So read on.

Nurses have specialized education and training validated by "professional licensure" in each state. We have a code of ethics and established practice standards we are bound to adhere to, a violation of which can result in our license being revoked or sanctioned. We have our own body of ongoing research that shapes and governs our practice. Nurses work autonomously without our scope of practice. We formulate and carry out our own plan of care for clients (when applicable); we apply judgment, use of critical thinking skills, and make nursing diagnosis.

Nurses use their specialized knowledge, experience, and skill set to initiate life-saving measures, improve and promote the health and well-being of the planet, and ease pain, suffering, and loss. We are all united in that common mission—regardless of where we work, our position title, or whether we're employed, unemployed, or self-employed.

Nurses whether directly or indirectly working with consumers of healthcare they have always been working within the profession of nursing. In each role the nurse had the same mission, ideals, and ethical and practice standards, while being aware of their role and responsibility as a healthcare expert (every nurse is a healthcare expert in his or her own way) and provider of care in a very broad sense. Today, as a nurse entrepreneur, when people ask me what I do, I say, "I am a self-employed registered nurse who spends her time speaking and writing. You might say I heal with words."

Since 1976, patient guidance has been an essential component of Registered Nurses' professional role. In today's healthcare delivery system, patient guidance and leading is becoming increasingly important. The public demands more information and control over their health care. Moreover, illness and limited knowledge about medicine, nursing and the healthcare system mean that patients are often vulnerable and powerless. Furthermore, the development of medicine and advanced technology has resulted in aggressive and swift use of technologies, which is becoming the norm in the contemporary hospital environment. In such a healthcare setting, patients' quality of life and right to self-determination tend to be ignored. In addition, inequalities and inconsistencies exist in the provision of healthcare resources at all levels in the United States of America (USA).

The philosophy of the nursing profession, nurses' educational background, and the unique position of nurses in the healthcare system mean that they should be able to guide and lead effectively for patients. The American Nurses Association (ANA) (2001) Code of Ethics for Nurses with Interpretive Statements requires that nurses guide the patients and lead them for, and protect the health, well-being, safety, values, and rights of patients in the healthcare system. This role is also reflected in the principal elements of the International Council of Nurses (ICN) (2006) Code of Ethics for Nurses:

In providing care, the nurse promotes an environment in which the human rights, values, customs and spiritual beliefs of the individual, family and community are respected. The nurse ensures that the individual receives sufficient information on which to base consent for care and related treatment.

The patient guidance role has been emphasized in the nursing profession. However, in reality, nurses have not put it fully into practice in the healthcare system, even though they may support the concept. The definition of patients' guidance is still confusing, and there is no consensus about its meaning among nurses and nurse authors. Neither the ANA (2001) Code of Ethics for Nurses nor the ICN (2006) Code of Ethics for Nurses contains a definition for the nursing profession. In the nursing literature, there is still a lack of understanding about nurses' views regarding guidance for patients and their actual performance of patient care. Most empirical studies related to nurses' patient promotion roles are qualitative, descriptive studies. Few quantitative studies on patient leading and guidance exist in the nursing literature. A clearly defined concept of patient guidance is necessary to develop quantitative research on patient guidance and leading. Our study therefore aims to clarify and refine the concept of patient guidance through synthesizing the guidance literature, and to establish a theoretical basis for future studies on patient guidance in nursing.

The Concept of Leading:

The Collins English Dictionary definition of the word leading is, "active support, (Collins English Dictionary, 22). Recognized that the concept of patients' leading arises in regulation, where the consults a client before a case comes to risk. However (Mallik) claims that patient leading by the nurses for the patients is distinctly different from other leading roles. Mallik (1997) further notes a difference in the structure of the support relationship in law and in nursing. Whilst in law the word promotion relates to a "calling to" and the establishment of a contract between the parties, in nursing the action tends to reflect more a "giving of" of one's help to an individual.

Vaartio and Kilpi (2004,) define the concept of leading is coming synonymly to the concept of advocacy as coming from the Latin "advocates", meaning one who is summoned to give evidence. Vaartio et al (2004, 705) synthesized three definitions of advocacy derived from the empirical research of seventeen research articles. They were; advocacy as motivated by the patients' right to information and self-determination; advocacy stemming from the patients' right to personal safety and advocacy as a philosophical principle in nursing. Advocacy as a right to information and self-determination is described as "proactive" by the authors and involves but is not limited to; assisting the patient to define their wishes; informing them about their illness; rights and treatment options. Advocacy stemming from the patients' right to personal safety is described as "reactive" and involves protecting a patient when their human rights are endangered. Vulnerable patients such as those with cognitive impairment or those under sedation may require an advocate.

Nursing patients' leading is a relatively modern idea, its inception being in the patient support movement of the 1970's (Hanks 2008). Its importance and prominence are reflected by its inclusion by various nursing bodies into their codes of ethics (Hanks 2008, 468. Mallik 1998,). Despite this, opinion is polarized as to the nature and extent of nursing promotion. The idea that patients require leader does not seem to be in dispute. What is contentious is whether or not nurses are in the ideal position to undertake such work or whether the practice of leading for the patient should be re-assigned to nursing's professional associations (Welchman et al. 2005,).

Studies in This Area:

In 2002, a paper published by Hewitt in the Journal of Advanced Nursing, aimed to critically review the arguments debating the role of the nurse patients' leader . Hewitt noted an imbalance in the quantity of empirical research into the concept of nursing leading with the majority of research concentrating on theory and concept (Hewitt 2002).

By synthesizing empirical research that provides concrete examples of the challenges nurses face in the field it is hoped to illuminate how the theory of nurse patients' leading translates into practice. Nursing support activities have received less coverage in the research literature than the concept itself (Vaartio et al. 2006,).

Another research involving British nurses in senior positions which aimed to investigate and elucidate the practical difficulties, barriers and problems that nurse encounter when leading their patients. And he has revealed that the practice is subject to contradictions and paradoxes and can cause inter-professional conflict within the health care system (Mallik 1998).

Merriam-Webster's Collegiate Dictionary (1998) defines patients' leading and guidance as 'the act or process of supporting', and a campaigner as 'one that pleads the cause of another. In the nursing literature, Florence Nightingale emphasized measures by which environmental factors can be manipulated to put patients in the best condition for nature to act upon them. Historically, the ideal nurse has been defined variously as healer, champion of the sick poor, parent-surrogate, physician-surrogate, contracted clinician, personal counselor and health educator (Gadow 1980). Since the 1970s, advocacy has increasingly been discussed as an essential component of nurses' professional role and many definitions have been proposed in the nursing literature. The most frequently discussed guidance models in the nursing literature are human guidance model, theory of existential guidance, and functional model of patient guidance. In 1989, Fowler added another model – social guidance – to the guidance literature. We found no systematic review of the concept of patient guidance in the literature.

Two nurse philosophers, Gadow and Curtin, thought that nursing ought to be defined philosophically rather than sociologically, that is, defined by the ideal nature and purpose of the nurse–patient relation rather than by a specific set of behaviors. According to Curtin (1979) human guidance model, healthcare professionals, patients or clients are all human beings, and it is this commonality that should form the basis of the relationship between nurses and patients/clients. In the field of human advocacy, nurses get to know patients and attend them as distinct and unique human beings. They must be sensitive to individuals and their reactions to those needs created by illness which may threaten the integrity of the person. Gadow (1980) theory of existential guidance is based on the principle that freedom of self-determination is the most fundamental and valuable human right. According to Gadow, nurses should help patients become clear about what they want to do by helping them discern and clarify their values in a particular situation, based on the principle of self-determination, so that they reach decisions expressing their reaffirmed values.

Kohnke (1982) proposed a functional model of patient advocacy, which is simpler and more pragmatic. Kohnke's central belief is that individuals have a right to self-determination. According to this model, advocacy involves informing patients and then supporting the decisions they make along with their right to make that decision. Informing means supplying the patient with the information needed to make informed choices.

Curtin (1979), Gadow (1980) and Kohnke (1982) advocacy models have the same basis in a belief in personal autonomy that means individuals are permitted personal liberty and freedom to determine their own actions. The philosophical basis for the advocacy concept in the nursing literature reflects the dominant values (e.g. humanity, individualism, freedom and autonomy) of US society. However, each of the three advocacy models has its own connotation and emphasis. Curtin's model is considered a humanistic model that emphasizes patients' benefits and nurses' humanity. Gadow's existential advocacy model and Kohnke's functional advocacy model both emphasize patients' self-determination and incline towards a legal advocacy model, although there are some differences between the two. Gadow stresses that nurses should help patients discern their own values and that the decisions made by patients should truly reflect those values. Nurses are more personally involved with patients' decisions. According to Kohnke, nurses should provide patients with information that is adequate to help them make their own decisions. What nurses need to do is to inform patients and support whatever decisions they make. Nurses are less personally involved with patients' decisions.

Fowler (1989) proposed the social guidance model. This model retains nurses' concerns about individual patients, yet advances them beyond institutional walls, and calls for participation in social criticism and social change. Social guidance calls attention to inequalities and inconsistencies in the provision of care at both the micro- and macro-allocation levels, and it insists on change. Social guidance is rooted in the concept of social justice. Applied to health care, social guidance means there should be equitable access to adequate nursing and care for all. This concept of social guidance attempts to bring 'what is' into conformity with 'what should be'. It tries to correct both clinical and social injustices that fail to respect patients as persons, their rights, their values, or their dignity. As a model, this form of guidance encompasses the values that the three guidance models described above contain, yet surpasses them by broadening the concerns of nurses beyond those of the immediate bedside.

If we want to talk about guidance and leading models, the beast model that has been correlated and linked by guidance as a concept is although the four advocacy models are frequently discussed in the nursing literature, each of them reflects solely on one aspect of patient advocacy. Moreover, all the four advocacy models fail to point out that nurses' patient advocacy behaviors are context-based: that is, nurses take different actions to advocate for patients in different clinical situations.

The History of Patients' Leading in Nursing

Nelson (1998) describes how Florence Nightingale's concerns for patient safety constitute acts of leading and support. This dedication to the patient has, however, sometimes lead to the nurse being in opposition to doctors. Snowball (1996) notes that it was not until 1973, that references to nurses maintaining loyalty and obedience to doctors was removed from the International Council of Nurses code Cultural changes in the 1960's and 1970's lead to nurse theorists such as Henderson (1960) claiming that nursing was becoming patient rather than institution lead. The upsurge in feminism and civil rights in the 70's in the USA spread to the United Kingdom and resulted in the birth of the debate regarding nurse-doctor-patient power relations (Snowball 1996, 68).

Patients' leading in nursing ethics has been discussed since first appearing in the literature in 1973 when it was added into the Professional Codes of the International Council of Nurses. Patient leading as a central nursing role was identified in the Code of Professional Conduct of the United Kingdom Central Council for Nursing, Midwifery and Health Visiting in 1992 (Hewitt 2002). According to Mallik (1997), the patient support movement has its roots in the United States, arising from the strong emphasis on human rights. A paucity in empirical literature from outside the United States was noted by Snowball in 1996 (Snowball 1996). The nursing profession in the United States has dominated the influence of the acceptance of the role of nurses as patient leading in the United Kingdom (Mallik 1997).

The Meaning of Guidance in Nursing:

Patient leading in nursing has been described as an "ethical ideal" (Davis et al. 2003). Also leading in nursing has been described as participating with the patient in determining the meaning of health, illness, suffering and dying ; providing information and supporting patients in their decisions; pleading the cause of a patient; protecting the patient from unnecessary worry ; disclosing negligence and misconduct and valuing, appraising and interceding (Vaartio et al. 2006, 282). Leading and support has further been defined to include the acts of so called "whistle blowing" that is, making known public, institutions or practices that are deemed unethical or negligent (Davis et al 2003).

Study in This Area:

Attempting has been made within nursing science to clarify the concepts of nursing patient leading. In 2007 Bu et al (2007) published a paper which aimed to "clarify and refine the concept of patient leading through synthesizing the leading literature" because they believed the concept of patient leading lacked a consistent definition. Their study synthesized 217 articles and three dissertations published between 1966 and 2006. From this data it is claimed that three core attributes of the concepts of leading emerge. They are; safeguarding the patient's autonomy, acting on behalf of patients, and championing social justice in the provision of health care. (Bu et al. 2006) These first two themes, it is suggested, are born from the theories of Curtin, Gadow and Kohnke. The last, the theory of social leading

Also from this study the researcher clarified that The first core attribute of safeguarding a patient's autonomy and rights, is concerned with actions which respect and promote a patient's self-determination. There are however two warning, patients must first be competent and secondly they must want to be involved in their healthcare and to be fully informed. This concept of patient leading can be described as being concerned with patients' legal rights. (Bu et al. 2006,) The second core attribute of nursing support as synthesized by Bu et al, was "acting on behalf of patients". This involves acting for patients who are unable to represent themselves or who do not wish to represent themselves. Patients who are unconscious would belong to this group.

The third concept was "championing social justice in the provision of health care". It is concerned with nurses actively striving to make changes to address inequalities and inconsistencies related to the provision of healthcare. Bu et al (2006, 104) also characterize the nature of encouragement as being on a micro social level or on a macro social level. By this they mean support actions that either concern an individual and their treatment; a micro social support intervention, or on a macro social level such as those interventions aimed at addressing social injustice in health care provision. (Bu et al. 2006).

According to our proposed theory, patient guidance is viewed as a process or strategy consisting of a series of specific actions for preserving, representing and/or safeguarding patients' rights, best interests and values in the healthcare system. Based on our concept analysis, patient guidance includes three broad core attributes: (1) safeguarding patients' autonomy; (2) acting on behalf of patients; and (3) championing social justice in the provision of health care. The number of citations where core attributes are discussed in the literature. The three broad core attributes occur explicitly or implicitly in the guidance literature and reflect underlying meanings of almost all the empirical referents of patient guidance or the specific patient guidance actions identified from the literature

Safeguarding Patients' Autonomy

The first core attribute represents a series of specific actions that respect and promote patients' self-determination under situations in which patients are competent and want to be involved in their own health care. There are two assumptions in this meaning. The first is that individuals have primary responsibility for their own health and healthcare professionals have a responsibility to respect and/or promote patients' health. Second, it is assumed that individuals are competent to make their own decisions and act on their own behalf though they may need information and assistance to do so. This core attribute emphasizes patients' legal rights in the healthcare system and reflects the implications of Gadow (1980) theory of existential advocacy and Kohnke (1982) functional model of patient advocacy.

Acting on Behalf of Patients

The second core attribute represents a series of specific actions that preserve and represent patients' values, benefits and rights in situations where patients are unable or do not wish to help and represent themselves. These situations may include unconsciousness, or choosing to have nurses act on their behalf. In such situations, advocacy means that nurses need to represent and defend patients when their rights and benefits are jeopardized. Advocates act as representatives, protectors, surrogates and delegates. This core attribute reflects the implication of Curtin (1979) human advocacy model.

These two core attributes are two important aspects of nurses' patient guidance role in two different types of clinical situation; they complement each other and do not conflict. They stress that nurses take different guidance actions for individual patients under different circumstances. They represent the patient guidance role at the micro social level.

Championing Social Justice in the Provision of Health Care:

The third core attribute of guidance is that of championing social justice in the provision of health care. The meaning of this core attribute is based on the ethics of justice (universal access to adequate nursing and health care) and reflects Fowler (1989) social advocacy model. It calls for nurses to actively strive for changes on behalf of individuals, communities and society as a whole, so that inequalities and inconsistencies are identified and corrected. In this meaning, nurses become social activists by being involved in issues pertaining to health, education and welfare for people in institutions, the community, or society, and address themselves to a redistribution of power and resources. This attribute represents the patient guidance role at the macro social level.

This mid-range theory of patient advocacy shows to be highlight the patient guidance and leading as the center of the healthcare system. It highlights patients' legal rights and best interests, and nurses' humanity and justice in the provision of health care. In addition, it recognizes that social and clinical circumstances can influence nurses' patient guidance behaviors along with philosophical ideas.

Leading acronym is: Loyalty, engagement, accountability, decision making.

The Need for Patient Support and Leading:

By acting as leading nurses are able to empower weak and vulnerable patients releasing them from discomfort and unnecessary treatments. Patients also require protection from acts of incompetence by health care professionals. (Vaartio et al. 2004, 705.) Mallik (1997, 131) notes that whilst historically patients have always been deemed to become vulnerable as a result of their physical condition, it is only recently that cultural conditions have resulted in this vulnerability as being seen to impact upon the patient's autonomy thus instigating a requirement to advocate (Mallik 1997, 131).

However, they are not only weak and vulnerable patients that require backer. But they are in danger of entering a process of "learned helplessness" as a result of an "omniscient and uninformative" doctor, resulting in the inability of the patient to speak for them (Hewitt 2002).

Bu et al (2006, 104) describe the kinds of events or incidents which instigate an support intervention on both the macro and micro social level and describe these as "antecedents" as they pre-exist the occurrence of advocacy. (Bu et al. 2006, 104) describes the imbalance in health status and access to healthcare between whites and minorities over the past 40 years as a macro social antecedent. On the micro social level, patient vulnerability is the most commonly cited condition in the literature requiring an support intervention.

Vulnerable patients may be those who are illiterate or do not fluently speak the language of the health care system in which they are being treated. Patients may be deemed vulnerable through a learning disability. Patients may also be considered vulnerable as a result of their physical condition or the anxiety it causes, such as those patients suffering from cancer. The ability of patients who are suffering mental illness or who are unconscious as a result of procedural intervention or accident are considered vulnerable in this respect. It has been noted that some patients who are otherwise competent in normal circumstances become "tongue tied ", shy and scared in the presence of the doctor. Other antecedents include patients who have been treated unethically, negligently or incompetently. (Bu et al. 2006).

ANTECEDENTS OF PATIENT GUIDANCE AND LEADING

Antecedents are those events or incidents that have a prior occurrence (Walker & Avant 1995). Antecedents of patient guidance and leading, occurring at the macrosocial and microsocial levels in the healthcare system.

Macrosocial Antecedents

On the macrosocial level, health disparity is one of the major antecedents of patient guidance and leading. This disparity often exists between white and minority populations. Research has consistently shown that, on almost any measure, minorities have poorer health than do whites, and such health disparity in the USA is increasing among. Factors contributing to this situation include poverty; access to, and use of healthcare services; individual and institutional racism; and, cultural difference between the biomedical health system and minority populations. Hospital environment is another antecedent. Use of advanced technology, healthcare costs and changing health policies mean that the hospital environment is becoming overwhelmingly complex (Donovan

1988). In such an environment, patients' autonomy and values can easily be ignored.

Micro Social Antecedents

On the micro social level, some patients' conditions are the major antecedents of patient guidance and leading. Patient defenselessness is the most frequently cited condition demanding nurses' patient guidance and leading actions. Vulnerable patients or populations refer to individuals or groups who cannot fully represent and protect their own rights, needs, benefits and wishes, are unable to make appropriate decisions, or are unable to carry out their decisions. Patients who are illiterate or lack command of the English language, which are unaware of their right to refuse treatment, or are unable to comprehend instructions or directions for some reason (e.g. learning disability), are vulnerable, and usually have difficulty in giving truly informed consent. Patients from lower socioeconomic groups or minority populations are vulnerable and tend to be underserved. Patients' illness conditions such as unconsciousness, cancer, and mental illness can cause vulnerability and compromise their ability to self-determine their health care and protect their best interests. Feelings of powerlessness because of limited knowledge about health care, or experience of being neglected in the system, can increase patients' vulnerability. In addition, the negative way that health professionals sometimes relate to patients, specifically the disregard, dehumanizing, controlling, punitive and judgmental practices of biomedicine, can lead to varying degrees of

21

susceptibility in patients (McCurdy 1997, Mitchell & Bournes 2000).

There are also other antecedents where nurses need to patient guidance and leading for patients. One is that patients are intimated under some circumstances. They may be fully autonomous under normal circumstances but become tongue-tied when their doctors come into the room, and are shy or scared to ask questions. There also situations where patients' rights and/or benefits are jeopardized. For example, a competent patient may wish to discontinue treatment but his or her physician or family disagrees with this wish. Other micro social antecedents are that patients are treated unethically or incompetently by some members in the healthcare team. In addition, sometimes patients verbally request nurses to act on their behalf (Segesten 1993).

CONSEQUENCES OF PATIENT GUIDANCE AND LEADING:

Consequences of patient guidance and leading are those events that occur as a result of nurses' patient guidance and leading actions, and can be either positive or negative. Corresponding to the attributes and antecedents of patient guidance and leading, consequences occur at both macro- and micro-social levels.

Positive Consequences

Nurses' successful patient guidance and leading actions produce positive consequences. At the micro social level, positive consequences mean that patients' rights, benefits and values are preserved or protected through nurses' particular advocacy actions. Patients are empowered and their autonomy is preserved. They get adequate and timely information regarding their health status and health care and so can make their own decisions. Patients can get prompt and appropriate treatments such as appropriate pain management. Successful patient advocacy can also improve patients' quality of life. And increase their safety in the health services. Nurses' patient guidance and leading actions may encourage patients to develop positive attributes such as a positive self-concept, personal satisfaction, self-efficacy, a sense of control, and a feeling of hope (Gibson 1991, Rushton 1994, Lindahl & Sandman 1998).

Successful patient guidance and leading may also produce positive effects on the nursing profession and nurse advocates (O'Connor & Kelly 2005), such as enhancing their public image and improving their professional status (Bernal 1992). By successfully sponsor for patients, nurses can increase their professional satisfaction, self-confidence and self-esteem, and maintain their personal integrity and moral principles.

Positive consequences at the macro social level means that nurses' patient guidance and leading actions lead to desirable changes for the well-being of a group of patients or society in general. Participating in policy-making is viewed as a type of nurses' patient guidance and leading. Changing inappropriate rules or policies in the healthcare system may promote social justice in the provision of health care and improve the quality of healthcare delivery, thereby enhancing patients' well-being. For example, DiGaudio reported in her study that health policies developed by a nurse actually improved the safety of school transportation for medically frail children. Participating in policy-making can not only increase nurses' credibility but also lead to other policy-making opportunities. Increased participation in policy-making can promote nurses' control over nursing practice and enhance the respect given to nurses as professionals.

Negative Consequences

In some situations, negative repercussions for individual backer s can occur. Risks are often reported and discussed when nurses advocate for patients. Patient advocates are sometimes accused of insubordination and suffer loss of reputation, friends, and self-esteem, or are labeled as trouble-makers or bad co-workers by nursing colleagues. also can be as advocates risks may experience extreme conflicts in the form of moral distress or moral dilemma, and feel powerless to do the right thing. Some very difficult situations where nurses practice advocacy may result in loss of professional security and lead to legal action. In some extreme cases, whistleblowers can encounter ostracism and experience severe disruption and havoc in their personal lives (Perrin 1992, O'Connor & Kelly 2005).

General Benefits and Consequences of Effective Patient Leading and Support:

The consequences of support& leading for the patients have only been reported as beneficial in contrast with those reported for nurses. For patients, positive benefits manifest as positive health outcomes. Vaartio et al (2004) report very specific patient outcomes such as increased patient survival in care of the elderly and increased birth weight of babies of low income mothers as positive effects of advocacy interventions. On a general level, positive consequences include preserving and protecting patient's rights, values and autonomy and empowering the patient. With regards to social justice support, participating in policy making and changing inappropriate rules are anticipated positive outcomes. Positive consequences for nurses include professional autonomy and proficiency (Vaartio et al, 2004). Bernal (1992) reports positive consequences including enhancing and improving the public image and professional status of nurses.

Risks to nurses from patient support and leaders are also discussed. In this area the majority of our authors affirmed that when the nurses try to lead their patient with some sort of support can suffer from some risks such as advocacy risks. The nurses stem largely from the conflict of loyalties and accountabilities of the nurse within the healthcare system. (Mallik 1997, 136). Nurses acting as advocates have been labeled as trouble makers by colleagues, accused of insubordination and have suffered the loss of reputation, friends and self esteem. Patient advocates may experience moral distress due to moral dilemma resulting in a feeling of powerlessness. Whistle blowing has been reported to result in ostracism and disruption extending to nurses personal lives. And the negative consequences for nurses include loss of job, status or professional role or direct conflict with the organization (Vaartio et al, 2004).

Despite the fact that nurses are obliged by their professional associations to lead their patients in their decisions, there remains little practical support and protection leaving the nurses potentially exposed to conflict. Salvage (1985) has written on the medical hostility attracted by nurse advocacy. Independent advocates have been suggested by Mallik (1997) and have become a reality in the United States of America and the United Kingdom.

Leading also has been considered as an active process of supporting a cause or position. However, patients' support has not always been a clear expectation in nursing. Seminal documents in the development of the American nursing curriculum, such as Nursing and Nursing Education in the United States and A Curriculum Guide for Schools of Nursing do not explicitly mention patient support or encouragement. Early nursing education emphasized conformity and a position subservient to the physician. Isabel Hampton Robb, an early leader in the development of American nursing education, encouraged obedience as the primary activity of the nurse. In 1900 Robb stated:

All the nurse remember to do what she is told to do, and no more; the sooner she learns this lesson, the easier her work will be for her, and the less likely she will be to fall under severe criticism. Implicit, unquestioning obedience is one of the first lessons a probationer must learn, for this is a quality that will be expected from her in her professional capacity for all future time (Hamric, 2000).

While Nightingale expected obedience in following the rules and medical direction, her intent was to allow nurses the autonomy of purpose to support the patients and the profession. It is probable that she would have disapproved of Robb's emphasis on obedience.

The term 'patient leading' was first utilized in the nursing literature by the International Council of Nurses in 1973. (Vaartio & Leino-Kilpi, 2004). Today the American Nurses Association (ANA) states that high quality practice includes guidance as an integral component of patient safety (ANA). Support is now identified both as a component of ethical nursing practice and as a philosophical principle underpinning the nursing profession and helping to assure the rights and safety of the patient. Nurses are seen as campaigner both when working to achieve desired patient outcomes and when patients are unable or unwilling to advocate for themselves.

Since 1973 advocacy has been considered a major component of nursing practice - politically, socially, professionally, and academically. Despite the seeming lack of a professional focus on encouragement before the early 1970s, it is argued that Nightingale implicitly laid the foundation for nurse support and established the expectation that nurses would advocate for their patients.

Every nurse has the opportunity to make a positive impact on the profession through day-to-day advocacy for nurses and the nursing profession. every nurse can employ to advocate for a safe and healthy work environment; and explains how nurses can advocate for nursing as part of their daily activity whether they are point-of-care nurses, nurse managers, or nurse educators. The advocacy practices discussed are applicable whether advocating on one's own behalf, for colleagues at the unit level, or for issues at the organizational or system level. (Benner2010)

Changes can challenge resource allocation decisions and adversely affect the work environment [and] can also create opportunities for nurses and the nursing profession. These are challenging times in which to be employed in healthcare. Unprecedented changes in the healthcare system are impacting care in all practice settings. These changes include financial pressures, uncertainty of the direction of healthcare reform, mandates from regulatory agencies to improve quality and patient safety, advancing technology, looming workforce shortages, and changes in the patient population. These changes can challenge resource allocation decisions and adversely affect the work environment. However, these forces can also create opportunities for nurses and the nursing profession. These opportunities include a greater voice for nursing in healthcare policy, expanded employment opportunities, and an enhanced image for nurses and the profession (2010).

In order to successfully capitalize on these emerging opportunities, it is important for nurses to work together, across employment settings and roles, to advocate on behalf of colleagues and the profession. Nurses comprise the largest professional group within healthcare and have been recognized by the public as the most trusted profession (Gallup, 2010; Jones, 2010). Despite nursing's strengths inherent in its size, diversity, and unique relationship with the public, the full potential for influence by the nursing profession has yet to be realized (Benner2010).

Although nurses anticipate future benefits resulting from healthcare system reform, the stress of today's workplace falls squarely on the shoulders of nurses at the point of care. To reap these future benefits, nurses need to advocate for the profession's desired future. It is important that all nurses engage in, and become involved in developing processes in their respective work settings to advocate for realistic changes that meet the needs of both patients and staff.

Patients' guidance often requires working through formal, decision-making bodies to achieve a desired outcome. Support is defined by the Merriam-Webster Collegiate Dictionary (2009) as the act or process of supporting a cause or proposal. An campaigner is defined as one that supports a cause or interest of another. Much of the literature on promotion comes from non-profit and special interest groups that prepare potential to influence public policy.

Study in This Area

Study of Hillman, A. (2005). Which aim was to discover the different Strategies promoted by these the following groups the study also applicable for nurses and the nursing profession. And stated that Amidei (2010) has described it as "seeing a need and finding a way to address it". Sharma (1997) defined as "action aimed at changing the policies, positions or programs of any type of institution".

Family Care International (2008) promoted as "the process of building support for an issue or cause and influencing others to take action"; while the Worldwide Palliative Care Alliance (2005) identified as "a process that can lead to change through influence" and a "way of directing decision-makers towards a solution". These definitions all suggest that the role of an advocate is to work on behalf of self and/or others to raise awareness of a concern and to promote solutions to the issue. Encouragement often requires working through formal, decision-making bodies to achieve a desired outcome. This process could include the 'chain of command' within a healthcare organization, a commission, a state legislature, or other groups at the healthcare system's policy level.

While most nurses readily embrace the mandate of the professional nurses' advocacy role as it applies to patients, the expectation for encouragement on behalf of colleagues, the profession, or even oneself may not be so clear or consistently noted. The professional responsibilities of the nurse to work with colleagues to promote safe practice environments are described in the American Nurses Association's (ANA) foundational documents, including the Nursing Scope and Standards of Practice (2010) and the Code of Ethics for Nurses with Interpretative Statements (Code of Ethics) (2001).

The ANA Scope and Standards of Practice identifies guidance for safe, effective practice environments as a responsibility of the professional nurse (ANA, 2010).The Code of Ethics describes the responsibility of the nurse to work through appropriate channels to address concerns about the healthcare environment. In addition, the Code of Ethics identifies a range of leading skills and activities that nurses are expected to demonstrate.

These activities promote the profession and form the basis of the guidance role for the professional nurse. The skills include service to the profession through teaching, mentoring, peer review, involvement in professional associations, community service, and knowledge development/dissemination (ANA, 2001). These activities and skills form the basis of guidance& leading role of the professional nurse.

PATIENTS' LEADING AND GUIDANCE SKILLS:

The ability to successfully support a cause or interest on one's own behalf or that of another requires a set of skills that include problem solving, communication, influence, and collaboration. Each of these skills will be discussed below.

Problem Solving:

It is important to take the time to develop a compelling request and to identify the appropriate time and individual to whom to make the request. Leading is focused on addressing problems or issues in need of a solution. The steps in the process are first to identify the issue(s) to be addressed and develop goals and a strategy to address the issue(s).

Once the strategy is identified, a plan of action is developed to organize an effort and establish a time line for completing each activity that supports the strategy. Most guidance initiatives involve approaching decision makers with requests for action to address the identified issue. Before approaching decision makers, however, it is important to take the time to develop a compelling request and to identify the appropriate time and individual to whom to make the request. Patience and a sense of timing are necessary in order to achieve a successful outcome. Few victories are achieved on the first attempt. Most guidance initiatives are accomplished through collaboration, negotiation, and compromise; they may require a series of actions over time in-order-to achieve a desired outcome (National Council of State Boards of Nursing, 2008).

Communication:

Successful leading requires effective communication skills. Most guidance& leading initiatives involve bringing individuals and groups together to address an issue or concern. Need to communicate clearly and concisely and to structure the message to fit both the situation and audience. It must be comfortable with verbal, written, and electronic formats. Communication regarding the issue should be factual and consistent. It is equally important to discuss the impact of the situation on those involved. It can be helpful to put a 'human face' on the issue by using 'word pictures' (words that create a picture in another's mind) to make the communication more compelling (Selanders, L. C. 2005).

One way to help to formulate a consistent communication message is to prepare a 'Sixty- Second Speech.' This is a brief, practiced speech used to introduce the issue and proposed solution. Distributing a one-page fact sheet or brochure is an excellent way to close the speech, and ensure that the listener is walking away with the key points (Amidei, 2010). The following (Box. 1) describes the content to include in a Sixty-Second Speech.

(Box 1) Items of a sixty-second speech to lead / guide patients
1- Your name, where you work or live, and department or agent where you are representing.
2- Describe the issue you are addressing.
3- Put a human face on your request, paint a word picture, and/or tell a story.
4- Describe what you would like the person/group to do.
5- Distribute a fact sheet describing your request and contact information.

Influence

Influence is built on competence, credibility, and trustworthiness. To facilitate change or solve an issue, the advocate must be able to influence others to action. Influence is the ability to alter or sway an individual's or group's thoughts, beliefs, or actions. Influence is built on competence, credibility, and trustworthiness. Keeping the best interests of those involved in the situation builds trust and credibility. An effective influences decision makers by building a case for the desired change, backing the case with facts and data, and putting a human face on the issue using a compelling visual image. Persuasion is a stronger form of influence that makes use of an appeal or argument to make one's point. While effective in small increments, persuasion can elicit defensiveness in others, thus undermining the overall success of an initiative. Matthews, J. H. (2010).

Collaboration

Collaboration is working with other individuals or groups to achieve a common goal. It differs from cooperation which involves groups working together to achieve their own individual goals. In addition to demonstrating the skills described above, the leader must also establish positive, collaborative relationships with others to garner the support necessary to address the issue. Collaboration is working with other individuals or groups to achieve a common goal. It differs from cooperation which involves groups working together to achieve their own individual goals. In collaboration, the individuals or groups involved develop common goals, along with common strategies and activities that will achieve that goal . Collaboration is built on trust, mutual respect, and credibility. The end result of groups collaborating to achieve a common goal can be greater than that which each group could accomplish independently. Successful collaboration requires careful communication with the groups involved in the process, seeking input when appropriate, and providing ongoing reports related to progress on achieving the goal. Donaldson, S. K., & Crowley D. M. (1978).

It is necessary, to work with those people (the stakeholders) who are affected by the issue. In addition, the campaigner may collaborate with others in the organization interested in solving the issue. These individuals often have expertise that would be beneficial to the effort. Developing a collaborative relationship with professionals in support departments, such as infection prevention, employee health, or human resources, will be invaluable when addressing issues that involve these departments. Likewise seeking out support staff in other venues, such as a legislative aid or the assistant to a commissioner, can be equally helpful (Gallagher, R. M. 2010)

Patient guidance and support has received international recognition over the past two decades. The concept has enjoyed greater acceptance and become an integral part of nursing practice. National nursing organizations in some countries (e.g. USA, UK, Australia and Canada) have included patient guidance and support as apart from patients' advocacy in codes of professional conduct. Despite this adoption of patient guidance and support internationally, people are still confused about the precise nature of the concept and what it means in practice. Early attempts to conceptualize patient guidance and support in nursing focused on philosophical definition of the concept by authors such as Curtin (1979) and Gadow (1980). Recently, nurse researchers such as Mallik (1997), Chafey *et al.* (1998), Lindahl and Sandman (1998), Breeding and Turner (2002), Davis *et al.* (2003), Carver and Morrison (2005), O'Connor and Kelly (2005) and McGrath and Walker (1999) have attempted to investigate further the role of nurses as patient advocates by conducting qualitative studies concentrating on nurses' experience and perceptions of patient advocacy. In contrast to the philosophical work on patient advocacy, these studies reveal various and dynamic expressions of the concept. Chafey *et al.* (1998) indicated that the nurse–patient relationship emerged as a salient feature of guidance and support, and teaching, informing and supporting were frequent activities of nurses in what they described as patient guidance and support.

Lindahl and Sandman (1998) described the nurse's role of patient guidance and support as building a caring relationship, carrying out a commitment, empowering, making room for and interconnecting, being a risk-taker and moral agent. These empirical studies suggest that the concept of patient advocacy is complicated and that there are different interpretations about patient advocacy among nurses and nurse researchers.

Inconsistency of interpreting the concept of patient guidance and support could be one of the major barriers for nurses' patient guidance and support practice and the advancement of research in the advocacy area.

Patients' Guidance& Leading System and Policies:

We have identified and categorized different precursors calling for nurses' patient guidance and support actions. If nurses know the antecedents well, it will be easy for them to judge whether patients need them to act as guidance and support and what strategies should be employed. For example, if nurses find some patients are fully autonomous under normal circumstances, but get tongue-tied when their doctors come into the room and are shy or scared to ask questions, they should help patients' communication with their doctors. In addition, the three core attributes of patient guidance and support in our proposed theory provide principles for advocating for patients in different situations. If patients are competent and want to be involved in their health care, nurses should respect and promote their autonomy. If patients are incompetent or want nurses or someone else to represent them, nurses should act on their behalf, representing their wishes or best interests. If some policies in the healthcare system generally interfere with patients' health care, nurses may be also need to take action to change inappropriate policies.

Knowing patients' best interests is very important for nurses when acting as patient guidance and support. Only when nurses know what patients' best interests are in a particular situation will patient guidance and support actions be meaningful, appropriate care provided and patients' needs satisfied. Communicating with patients and/or their families is important in order to understand their best interests. Communication is, however, a two-way process and, where appropriate, nurses also need to provide relevant information about patients' health situation (e.g. diagnosis, healthcare options and prognosis) and make sure that patients are fully informed so that they can make decisions consistent with their best interests.

It is critical that nurses act as patient guidance and support. If it is known that a patient needs a guidance and support and no action is taken, then the appropriate help and care will not be received. It is important for healthcare administrators to motivate nurses to take actions to guidance and support for patients based on their knowledge of patient advocacy. If nurses lack knowledge and skills related to patient guidance and support, education or training may be necessary. For example, nurses may need to know how to negotiate between patients, families and other healthcare providers (e.g. physicians). Good

Among the challenges confronting patients with rare diseases is a dearth of treatment options. The development of safe and effective new therapies is hampered by challenges associated with conducting clinical trials in small populations. a proposed draft guidance document for organization for submission to the U.S. Food and Drug Administration. This unprecedented undertaking involved a broad coalition of more than 80 stakeholders collaborating across nine time zones to produce a document in only 6 months. We hope that other rare disease communities and advocacy organizations can use our experience as a model for developing their own draft guidance documents.

Several factors need to be in place before a patient guidance group community should consider developing draft guidance. First, the disease must be characterized. Areas of unmet need should be established and discussed with leading experts in the field, and the FDA. Working as a partner with the FDA, PPMD has conducted studies10 to provide the data needed to inform decisions. Developing these relationships as partners in the fight against different disease were critical to successfully developing this guidance

Evidence Base and Patients' Guidance:

As is typical for evidence-based reviews, the goal was to provide a critical appraisal of the evidence on the topic. This information would then be available to others to ensure that no practice unsupported by evidence would be endorsed and that no practice substantiated by a high level of proof would lack endorsement. Readers familiar with the state of the evidence regarding quality improvement in areas of health care where this has been a research priority (e.g., cardiovascular care) may be surprised and even disappointed, by the paucity of high quality evidence in other areas of health care for many patient safety practices. One reason for this is the relative youth of the field. Just as there had been little public recognition of the risks of health care prior to the first IOM report, there has been relatively little attention paid to such risks – and strategies to mitigate them – among health professionals and researchers.

Moreover, there are a number of methodological reasons why research in patient guidance is particularly challenging. Cannot be the subject of double-blind studies because their use is evident to the participants, Second, capturing all relevant outcomes, including "near misses" and actual harm, is often very difficult. Third, many effective practices are multidimensional, and sorting out precisely which part of the intervention works is often quite challenging. Fourth, many of the patient guidance problems that generate the most concern are uncommon enough that demonstrating the success of "guidance practice" in a statistically meaningful manner with respect to outcomes is all but impossible.

Points of Patients' Care as Contributing to Nursing Profession:

Never before has the voice of the nurse at the bedside been so critical. The impact of leading nurses on patient outcomes is increasingly evident. It is essential that point-of-care nurses develop and use advocacy skills to address workplace concerns, promote positive work environments, and advocate for the profession. Never before has the voice of the nurse at the bedside been so critical to patients, colleagues, and healthcare facilities. An increasing number of facilities have, or are developing shared governance structures to ensure that nurses at the point of care have a voice in decisions related to patient care and the work environment.

The impact the nurses on the patient's outcomes is increasingly evident; and nursing input into organizational decision making related to safety and quality initiatives is invaluable. Nurses are increasingly positioned to advocate more effectively than ever before not only for patients, but also for themselves and the nursing profession. International Council of Nurses. (2011)

Opportunities for Patients' Care Leading and Guidance:

The outcome for point of leading patients' Care can be very significant for improving working conditions for all staff. Membership on committees, councils, and quality improvement teams provides opportunities to advocate. When serving on a committee, council, or team, it is important to represent the needs of both colleagues and patients. Sometimes this means considering the impact of an issue or proposed solution on nurses and staff in other departments as well as one's own workgroup. The best way to work through the needs of multiple groups is to consider what ultimately is best for the patient, client, or population served.

Engagement in organization-wide activities provides opportunities to advocate for colleagues and for the profession. Many organizations conduct periodic, employee satisfaction or opinion surveys that are used to develop plans to promote staff engagement. While the time an employee invests in completing a survey may be only a few minutes, the outcome can be very significant for improving working conditions for all staff. MacIver, R. M. (1955).

Modeling positive professional behaviors and helping those new to the profession to acquire these behaviors is a form of advocacy. Nurses have an opportunity for advocacy when involved in teaching nursing students and new nurses at the bedside. Students and new nurses are excited about the profession they have chosen. They see practicing nurses as role models and mentors. Modeling positive professional behaviors and helping those new to the profession to acquire these behaviors is a form of advocacy. Providing guidance during a difficult learning situation, such as the first time a novice performs a procedure, can advocate for both the patient and the novice.

Opportunities occur at many levels: some occur in the work setting and others may occur in the grocery store. In another agency nurses were concerned about the increasing incidence of back injuries among the nursing staff. The staff approached the hospital risk manager who organized a task force to develop a program to reduce back injuries. Nurses, nursing assistants, physical therapists, and transporters were all involved in developing the program and testing products. They reviewed the lift and transfer devices available to facilitate safe patient handling and ensure staff safety. In addition, they assisted with training on the use of the equipment, which over time included ceiling-mounted lifts and transfer devices. The committee members also served as champions for eliminating manual patient lifting. As a result, the incidence of staff injuries decreased significantly. Matthews, J. H. (2010).

Every nurse can play a role in advocating for nurses and the profession. It is through day-to-day collective action that nurses work together to advocate for improvements in the work environment and for the advancement of the profession. Opportunities for advocacy occur at many levels: some occur in the work setting and others may occur in the grocery store. The key is to promote the profession with every opportunity that arises.

MANAGER ROLE IN LEADING PATIENTS' CARE:

Leaders direct& guide for patients, nurses, and the profession in a number of ways. This can include actions both to ensure appropriate resource allocation and to promote positive work environments. Merton, R. K. (1958).

Leading an Effective Work Environment

Today's work environment is increasingly stressful, and competition for resources is keen. Nursing leaders can advocate for staff by actively involving staff in decisions that directly affect the practice environment. Guidance is enhanced when scheduling and staffing are a collaborative process that involves staffing committees and self-scheduling approaches. Staff involvement can help to ensure balanced schedules and flexible staffing approaches that meet the needs of both patients and staff. In addition, proactive planning to formulate solutions to unpredicted staff shortages can facilitate patient and staff safety in unforeseen situations

Leaders also fulfill the advocacy role by protecting nursing resources during times of budget scrutiny, work process redesign, or work flow change. Staff involvement in the budgeting process promotes an understanding of the challenges operating in today's healthcare environment. Staff can be included in a number of ways, for example by providing input on and prioritization of equipment and supply purchases. Increased staff knowledge of the costs associated with procedures also promotes effective usage and cost containment. When staff is involved in organizational initiatives, they are more likely to promoter for, and foster adoption. Collaboration between nursing managers/administrators and staff nurses is essential for maintaining adequate resources. Merton, R. K. (1958).

Leading Effective Communication

When leaders support open communication, collaboration, and conflict resolution skills, staff are able to advocate more effectively for themselves and for colleagues. Managers play a pivotal role in developing the advocacy capabilities of staff. When leaders support open communication, collaboration, and conflict resolution skills, staff are able to advocate more effectively for themselves and for colleagues. In contrast conflict undermines effective teamwork and put at risk patient safety. Much has been written about the negative consequences of nurse incivility (Bartholomew, 2006; Longo, 2010). Fostering the development of conflict resolution skills and addressing unprofessional behavior, including incivility, promotes an environment in which promotion can flourish.

Leaders promote encouragement when they enable staff to autonomously address concerns. They foster staff ownership of issues when they refer a concern to staff councils and form task forces, involving other departments as appropriate. In such situations the role of the leader becomes primarily a coach who provides guidance, helps staff navigate within the organization, and removes barriers to the process. (National association of colored graduate nurses records, 1908)

Leaders promote support when they enable staff to autonomously address concerns. The hospital recruitment and retention committee, comprised of staff from a variety of nursing units, plus recruiters, staff development educators, and human resource professionals, met regularly to plan and evaluate recruitment and retention programs. The committee had already implemented a comprehensive nurse retention program that included recognition for national certification, incentives for nurse preceptors, and strategies to improve communication between nurses and physicians. One staff nurse on the committee felt that recruitment and retention could also be improved by providing an on-campus RN-to-BSN program. Prior to approaching the committee with this idea, he talked with nurses from across the organization to determine the level of interest and the program features that would accommodate working nurses. When he presented the idea to the recruitment and retention committee, he was able to identify the potential number of nurses interested in the program and volunteered to serve on a planning committee.

Communication skills are important for nurses to achieve this. The literature shows that there are many factors impeding nurses from taking action to guidance and support for patients; it is important for administrators to identify such barriers and minimize or eliminate them in order to facilitate nurses' patient guidance and support roles. One of the major barriers is a lack of support from institutions, colleagues, and/or administrators (Perrin 1992, Schroeter 2000, Svedberg *et al.* 2000). In addition, patients' guidance and support such as or considered as apart from patient advocacy actions sometimes carry negative consequences for nurse advocates which may discourage them from taking action; support from administrators could help to overcome this barrier. It is, therefore, important to provide support for nurses' patient advocacy actions and establish a supportive environment within the institution; ideally, administrators should actively encourage nurses to advocate for patients and reward their behavior.

Nurse Educator's Role in leading patients' Care

Nurse educators in professional development roles serve as the culture carriers for the profession. These educators are essential in the formation and continued development of nurses' professional identity as, an identity that transcends their entire career. One trend in healthcare over the past twenty years has been the active involvement of the nursing staff in decision making. This involvement increases the need for staff with more fully developed leadership skills and the ability to advocate effectively. No one plays a more critical role in developing the capacity and capability for professional advocacy than do nursing educators who model advocacy behaviors for students in both education and practice settings. Nurses in staff development roles contribute to this process of role formation by providing ongoing mentoring to nurses in practice. In many ways faculty in academic settings and nurse educators in professional development roles serve as the culture carriers for the profession. These educators are pivotal in the formation and continued development of nurses' professional identity as advocates, an identity that transcends their entire) career.(Bucher, R., & Strauss, A. (1961).

Educators involved in forming the professional identity of nursing students and shaping the capabilities of the nursing workforce are pivotal to advancing the profession. Healthcare is changing. Achieving the best possible future requires that nurses be prepared to advocate for nursing and for their professional roles. American Nurses Association. (2010).

Professional Organizations and Leading Patients' Care:

Professional organizations and associations in nursing are critical for generating the energy, flow of ideas, and proactive work needed to maintain a healthy profession that leading patients' Care for the needs of its clients and nurses, and the trust of society (Matthews, J., January 31, 2012).

Early on, certain individuals within each society began providing care and nourishment for those who were unable to care for themselves. As these individuals became 'care experts,' they began to share with others the practices that worked for them and to train others as apprentices who would carry on their work. The evolution of modern nursing from a vocation, to the discipline and profession of nursing, began in the late 1800s as Nightingale articulated her views about how nurses should be trained and educated and how patient care should be provided (Hegge, 2011).

Study in This Area

Study which conducted by (ANA, 2001), which the purpose was to describe the role of professional nursing organizations in supporting for the nursing profession and for nurses. He discussed the characteristics of a profession, review the history of professional nursing organizations, and describe the advocacy activities of professional nursing organizations. Throughout, explain how the three foundational documents of the nursing profession emphasize nursing advocacy by the professional organizations as outlined in the Code of Ethics for Nurses.

Patients' Care Leading Activities as A Professional Nursing Organizations

Leading is the cornerstone of nursing – nurses advocate for patients, causes, and the profession. Our advocacy, motivated by moral and ethical principles, seeks to influence policies by pleading or arguing within political, economic, and social systems, and also institutions, for an idea or cause that can lead to decisions in resource allocation that promote nurses, nursing, and all of healthcare.

Patients' leading Support as a from the Code of Ethics

Patients' leading by the profession of nursing developed within the US as visionaries, leaders, and nurses from across the nation formulated the first (and subsequent) revisions of the Code of Ethics for Nurses with Interpretive Statements. These codes of ethics about patients' leading listed in the (Box 2).

He added the '(patients' leading)' notation because the focus of this article is on nursing patients' leading for the nine provisions of the Code of Ethics for Nurses. In the Code of Ethics for Nurses, the concept of guidance and leading for the individual nurse is openly named in Provision Three (Box.2).

Box 2: Provisions of the Code of Ethics for Nurses, (ANA, 2001)

Provision	Statement
Provision 1	The nurse, in all professional relationships, practices with compassion and respect for the inherent dignity, worth, and uniqueness of every individual, unrestricted by considerations of social or economic status, personal attributes, or the nature of the health problems.
Provision 2	The nurse's primary commitment is to the patient, whether an individual, family, group, or community.
Provision 3	The nurse promotes, guide, lead and advocates for, and strives to protect the health, safety, and rights of the patient.
Provision 4	The nurse is responsible and accountable for individual nursing practice and determines the appropriate delegation of tasks consistent with the nurse's obligation to provide optimum patient care.
Provision 5	The nurse owes the same duties to self as to others, including the responsibility to preserve integrity and safety, to maintain competence, and to continue personal and professional growth.
Provision 6	The nurse participates in establishing, maintaining, and improving healthcare environments and conditions of employment conducive to the provision of quality health care and consistent with the values of the profession through individual and collective action.
Provision 7	The nurse participates in the advancement of the profession through contributions to practice, education, administration, and knowledge development.
Provision 8	The nurse collaborates with other health professionals and the public in promoting community, national, and international efforts to meet health needs (p. 23).
Provision 9	The profession of nursing, as represented by associations and their members, is responsible for articulating nursing values, for maintaining the integrity of the profession and its practice, and for shaping social policy (p. 24).

The Profession's leading Efforts:

In Provision 9 the professional associations, created by nurses for nurses to articulate nursing values, integrity, practice, and social policy, demonstrate advocacy and self-regulation. In the US, ANA is the organization that solicits and coordinates ideas from individuals, and from the nursing specialties and associations, deliberates regarding these ideas, and develops them based on the Code of Ethics and the other two 'framework documents' that serve as the basis of the nursing profession. The three framework documents include:

- The Code of Ethics for Nurses - asserts the values and commitment to excellence for patients, society, and nurses individually and collectively as a profession (ANA, 2001);
- The Social Policy Statement - details the authority, based on the social responsibility of the profession to society. It serves as nursing's contract between the profession of nursing and society to uphold the highest values and standards in delivering its service of nursing care (ANA, 2010); and
- The Scope and Standards of Practice in Nursing - delineates the scope of nursing practice and then defines the standards of professional nursing practice and accompanying competencies (ANA, 2010).

Our various nursing professional organizations can work together to advocate for nurses and nursing. This occurs by maintaining a spirit of unity, engaging in political advocacy, keeping nurses informed, disseminating professional knowledge, and promoting professional development.

Many specialty organizations...educate the public, policy makers, healthcare administrators, and professionals on specific issues. Each of the specialty organizations advocates for nurses as their organizational goals pertain to its members, specialty, and practice settings. Many specialty organizations, and their members, educate the public, policy makers, healthcare administrators, and professionals on specific issues. Nursing organizations are recognizant of the power of unity and engage in collaborative ventures with other nursing and health-related professional organizations when appropriate.

Patients' Care Leading and Guidance and Participating in Different Union Association

Nurses can respond in a variety of ways to ask decision makers to support and guide for patients. The ability of professional organizations to communicate quickly with their members is one of the many benefits of involving a variety of organizations in collaborative efforts. Newsletters and bulletin alerts keep members aware of issues and help explain developments that may affect nurses and patient care delivery. Issue-specific communication to members often request nurses to respond to late-breaking developments.

In this age of announcement, nurses can respond in a variety of ways, e.g. through phone calls, email, Tweets, and Face book™ postings, to ask decision makers to support and advocate for nurses, letting them know how a given proposal will affect those who give and those who receive healthcare. Of the more than three million nurses in the U S, 2.6 million are actively involved in the workforce many, if not most of them, have access to electronic communications. These nurses have the ability to analyze the information provided and to respond quickly. The power of over two million voices at the national level is awesome! It can significantly influence the development of policy and legislation (Bureau of Health Professionals, 2011).

The privilege of participating in association leading and patients' guidance is an important benefit of membership in one's professional organization(s). Today professional organizations have list-serves and networking information-sharing features that strengthen a nurse's ability to guide for nurses and nursing. The privilege of participating in association advocacy is an important benefit of membership in one's professional organization(s). As the associations continue to strengthen communication support structures, professional membership increases in value by offering a mechanism for communicating via secure, intra-member, social networks

To facilitate identification and evaluation of potential patient safety practices, the Editorial Board divided the content for the project into different domains. Some cover "content areas," including traditional clinical areas such as adverse drug events, nosocomial infections, and complications of surgery, but also less traditional areas such as fatigue and information transfer. Other domains consist of practices drawn from broad (primarily nonmedical) disciplines likely to contain promising approaches to improving patient safety (e.g., information technology, human factors research, organizational theory).

Once this list was created—with significant input from patient safety experts, clinician–researchers, AHRQ, and the NQF Safe Practices Committee—the editors selected teams of authors with expertise in the relevant subject matter and/or familiarity with the techniques of evidence-based review and technology appraisal. The researches were given explicit instructions regarding search strategies for identifying safety practices for evaluation (including explicit inclusion and exclusion criteria) and criteria for assessing each practice's level of evidence for efficacy or effectiveness in terms of study design and study outcomes.

Some safety practices did not meet the inclusion criteria because of the paucity of evidence regarding efficacy or effectiveness but were included in the report because an informed reader might reasonably expect them to be evaluated or because of the depth of public and professional interest in them. For such high profile topics (such as bar coding to prevent misidentifications), the researchers tried to fairly present the practice's background, the experience with the practice thus far, and the evidence (and gaps in the evidence) regarding the practice's value. For each practice, authors were instructed to research the literature for information on:

• prevalence of the problem targeted by the practice;

• severity of the problem targeted by the practice;

• the current utilization of the practice;

• evidence on efficacy and/or effectiveness of the practice;

• the practice's potential for harm;

• data on cost, if available; and implementation issues.

PATIENTS' LEADING MODELS:

There are many models of theories that linked with patients' care leading, the most famous of them were

1- Systems Theory Model:

This diagram of system model nursing factors the most influencing factor, a key player in each of the steps. Thus it should be obvious to you that nurses contribute and control, to a large degree, to the viability of their organizations This example is a very simplistic model of how nurses and patients interact within a health care system. Of course we know that within each component of this system a multitude of variables affects the nurse: patient interaction (Fig. 1).

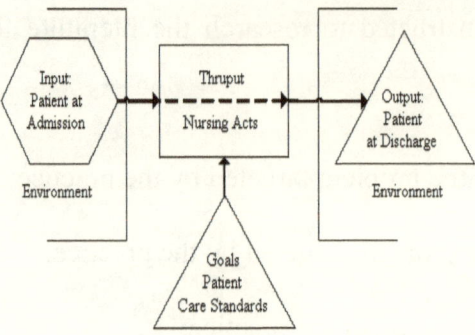

Figure 1: Basic Systems Theory model.

2-Model of advocacy theory which developed to help in dividing the role advocacy plays in the nurse: patient relationship in a busy Emergency department (fig. 2).

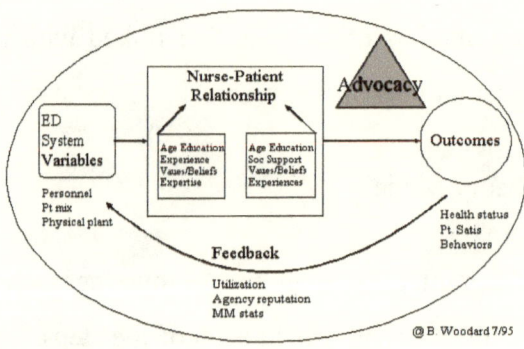

Figure 2: Model of advocacy theory.

As illustrated in (Fig. 3) the foundation for nursing underpins state nurse practice acts and institutional policies and procedures. The apex of this model of professional responsibility is achieved with the individual nurse's assumption of personal accountability for continuing education and professional experience over and above the basic requirements for professional nursing.(ANA2010).

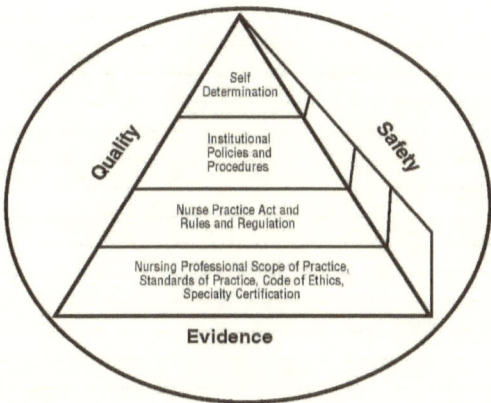

1- Figure 3: Professional nursing.

Summary:

Nurses work to promote health, prevent disease and help patients to cope with illness. They are health educators for patients, families and communities. When they provide a patient care, they assess, record and react and work in partnership with the patient. The provision of direct patient care in convalescence and rehabilitation. Nurses work in an environment that is constantly changing to provide the best possible care for patients. They are learning about the latest technology and medication as well as considering the evidence that their nursing practice is based upon. Because they will actually spend more face-to-face time with a patient than nurses must be particularly skilled at interacting with patients, putting them at ease, and assisting them in their recovery. It is often said that nurses care, so that patient guidance has been an essential component of the nurses' professional role.

Notes

References:

American Nurses Association. (2001). The code of ethics for nurses with interpretive statements. Washington, DC: Nursesbooks.org.

American Nurses Association. (2010a). Nursing's social policy statement: The essence of the profession (3rd ed.). Silver Spring, MD: Nursebooks.org.

American Nurses Association. (2011b). Congress of nursing practice and economics. Retrieved June 19, 2011 from www.nursingworld.org/FunctionalMenuCategories/AboutANA/Leadership-Governance/NewCNPE.

Attewell. Florence Nightingale today: Healing, leadership, global action (pp.66-74). Silver Spring, MD: American Nurses Association.

Bettany-Saltikov J. 2010 Learning how to undertake a systematic review: part 2.Nursing Standard Vol.24 No.51, 47-56.

Bu, X. & Jezewski, M. 2006. Developing a mid-range theory of patient advocacy through concept analysis. Journal of Advanced Nursing Vol.57 No.1, 101-110.

Bucher, R., & Strauss, A. (1961). Professions in process. American Journal of Sociology, 66 (4), 325-334. Available: www.jstor.org/stable/2773729.

Bureau of Health Professions. (2011). The registered nurse population: Findings from the 2008 National Sample Survey of Registered Nurses. Retrieved June 19, 2011 from http://bhpr.hrsa.gov/healthworkforce/rnsurvey2008.html.

Collins English Dictionary. Fourth Edition.

Davis, A.; Konishi, E. & Tashiro, M. 2003. A pilot study of selected Japanese Nuurses' ideas on patient advocacy. Nursing Ethics Vol. 10 No.4, 404-410.

Donaldson, S. K., & Crowley D. M. (1978). The discipline of nursing. Nursing Outlook, 33(1), 113-120.

Gallagher, R. M. (2010). Quality is NOT an irreconcilable difference: The patient protection and affordable care act (P.L. 11-148). Nursing Management, 41(8), 18-20. DOI 10.1097/01.NUMA.0000384004.09214.23

Hanks, R. 2008. The lived experience of nursing advocacy. Nursing Ethics Vol 15. No. 4, 468-477.

Hegge, M. J. (2011). The lingering presence of Florence Nightingale. Nursing Science Quarterly, 24(2), 152-162. DOI: 10.1177/0894318411399453.

Hewitt, J. 2002. A critical review of the arguments debating the role of the nurse advocate. Journal of Advanced Nursing 2002 Vol.37 No.5, 439-445.

Hillman, A. (2005). Reflections on service orientations, community, and professions. Academy of Management Journal, 48(2), 185-188.

International Council of Nurses. (2011) About ICN. Retrieved November 1, 2011 from www.icn.ch/about-icn/about-icn/.

Mallik, M. 1997. Advocacy in nursing – a review of the literature. Journal of advanced nursing Vol.25, 130-138.

Mallik, M. 1997. Advocacy in nursing – a review of the literature. Journal of advanced nursing Vol.25, 130-138.

Mallik, M. 1998. Advocacy in nursing: perceptions and attitudes of the nursing elite in the United Kingdom. Journal of advanced Nursing Vol. 28 No. 5, 1001-1011.

Matthews, J. H. (2010). When does delegating make you a supervisor? OJIN: The Online Journal of Issues in Nursing, 15(2), DOI: 10.3912/OJIN.Vol15No02Man03.

Matthews, J. H. (2010). When does delegating make you a supervisor? OJIN: The Online Journal of Issues in Nursing, 15(2), DOI: 10.3912/OJIN.Vol15No02Man03.

Matthews, J. H. (2010). When does delegating make you a supervisor? OJIN: The Online Journal of Issues in Nursing, 15(2), DOI: 10.3912/OJIN.Vol15No02Man03.

Matthews, J., (January 31, 2012) "Role of Professional Organizations in Advocating for the Nursing Profession" OJIN: The Online Journal of Issues in Nursing Vol. 17, No. 1.

Merton, R. K. (1958). The functions of the professional association, American Journal of Nursing, 58(1), 50-54.

Merton, R. K. (1958). The functions of the professional association, American Journal of Nursing, 58(1), 50-54.

National association of colored graduate nurses records, 1908-1951. History. (1984). New York Public Library Digital Library Collections. Available: http://digilib.nypl.org/dynaweb/ead/scm/scmnacgn/@Generic__BookTextView/143#X

National Council of State Boards of Nursing (2008). APRN model act/rules and regulations. Retrieved November 1, 2011 from https://www.ncsbn.org/APRN_leg_language_approved_8_08.pdf

Nelson, S. & McGillion, M. (1998) Expertise or performance? Questionaing the rhetoric of contemporary narrative use in nursing. Journal of advanced nursing, 47(6), 631-638

Nightingale's foundational philosophy for nursing. In D.B. Dossey, L.C. Selanders, D.M. Beck & A. Attewell. Florence Nightingale today: Healing, leadership, global action (pp.66-74). Silver Spring, MD: American Nurses Association.

Selanders, L. C. (2005a). Nightingale's foundational philosophy for nursing. In D.B. Dossey, L.C. Selanders, D.M. Beck & A.

Snowball, J. 1996. Asking nurses about advocating for patients: "reactive" and "proactive" accounts. Journal of Advanced Nursing Vol.24, 67-75.

The social significance of professional ethics. Annals of the American Academy of Political and Social Science, 297, 118-124. Available: www.jstor.org/stable/1029847.

Tomajan, K. (January 31, 2012) "Advocating for Nurses and Nursing" OJIN: The Online Journal of Issues in Nursing Vol. 17, No. 1, Manuscript 4.

Vaartio, H. & Leino-Kilpi. 2004. Nursing advocacy – a review of the empirical research 1990-2003. International Journal of Nursing Studies Vol.42, 705-714.

Vaartio, H.; Leino-Kilpi, H.; Salanterä, S.; & Suominen, T. 2006. Nursing advocacy: how is it defined by patients and nurses, what does it involve and how is it experienced? Scandinavian Journal of Caring Sciences Vol. 20, 282-292.

Welchman, J. & Griener G. 2005. Patient Advocacy and professional associations :individual and collective responsibilities. Nursing Ethics Vol 12. No. 3, 296-304.

Yahia, A., Ali, N. S., & Elhabashy, S. (2013). Factors Affecting Validity of Arterial Blood Gases Results among Critically Ill Patients: Nursing Perspectives. *Journal of Education and Practice*, *4*(15), 43-56.